Standard of Ministry of Water Resources of
the People's Republic of China

SL 168—2012
Replace SL 168—96

Code for Acceptance of Small Hydropower Station Construction

Drafted by:
National Research Institute for Rural Electrification of MWR
Translated by:
National Research Institute for Rural Electrification of MWR

China Water & Power Press
Beijing 2015

图书在版编目（CIP）数据

小型水电站建设工程验收规程 : SL168-2012 = Code for Acceptance of Small Hydropower Station Construction:SL168-2012 : 英文 / 中华人民共和国水利部发布. -- 北京 : 中国水利水电出版社, 2015.4
ISBN 978-7-5170-3150-5

Ⅰ. ①小… Ⅱ. ①中… Ⅲ. ①水力发电站－工程验收－规程－英文 Ⅳ. ①TV742-65

中国版本图书馆CIP数据核字(2015)第092945号

书　　名	**Code for Acceptance of Small Hydropower Station Construction SL 168—2012**
作　　者	中华人民共和国水利部　发布
出版发行	中国水利水电出版社 （北京市海淀区玉渊潭南路1号D座　100038） 网址：www.waterpub.com.cn E-mail：sales@waterpub.com.cn 电话：(010) 68367658（发行部）
经　　售	北京科水图书销售中心（零售） 电话：(010) 88383994、63202643、68545874 全国各地新华书店和相关出版物销售网点
排　　版	中国水利水电出版社微机排版中心
印　　刷	北京瑞斯通印务发展有限公司
规　　格	140mm×203mm　32开本　3.125印张　110千字
版　　次	2015年4月第1版　2015年4月第1次印刷
定　　价	**188.00**元

凡购买我社图书，如有缺页、倒页、脱页的，本社发行部负责调换

Introduction to English Version

Department of International Cooperation, Science and Technology of Ministry of Water Resources, P. R. China has the mandate of managing the formulation and revision of water technology standards in China.

Translation of this code from Chinese into English was organized by Department of International Cooperation, Science and Technology of Ministry of Water Resources, P. R. China in accordance with due procedures and regulations applicable in the country.

This English version of code is identical to its Chinese original SL 168—2012 *Code for Acceptance of Small Hydropower Station Construction*, which was formulated and revised under the auspices of Department of International Cooperation, Science and Technology of Ministry of Water Resources, P. R. China.

Translation of this code is undertaken by National Research Institute for Rural Electrification.

Translation task force includes Li Zhiwu and Liu Ruoxin.

This code is reviewed by Gu Liya, Guo Jun, Qiao Shishan, Meng Zhimin, Sun Feng, Jin Hai, Wang Yicheng, and Wu Suhua.

Department of International Cooperation, Science and Technology
Ministry of Water Resources, P. R. China

Foreword

According to the plan to formulate or revise technical standards of the Ministry of Water Resources of the People's Republic of China and SL 1—2002 *Compilation Requirements for Technical Standards of Ministry of Water Resources*, SL 168—96 *Code for Acceptance of Small Hydropower Station Construction* (hereinafter referred to as the previous code) is revised.

This code comprises 9 chapters, 12 sections, 113 articles and 19 appendices with the main contents as follows:

—Classification of acceptance.

—Supervision and management.

—Organizing and procedures for different types of acceptances.

—Essential conditions for different types of acceptances.

—Main contents and output files of all acceptances.

—Preparation of required reports and documents of acceptance.

—Handover of approved projects and treatment of remaining problems.

Revisions to the previous code are as follows:

—Adjusting the applicable scope of the code.

—Reclassifying acceptance names.

—Adjusting the provisions of the management and organizing of acceptance.

—Adjusting the structure of the code.

—Adding the following chapters:

—Supervision and management of acceptance.

—Acceptance of sub - unit projects.

—Acceptance of unit projects.

—Acceptance of contracted project completion.

—Acceptance of special item.

—Project handover and handling of leftover problems.

—Adjusting the contents of stage acceptance and addition of the previous acceptance of turbine-generator startup to the stage acceptance.

—Adjusting the contents of completion acceptance, cancellation of preliminary acceptance and addition of self-inspection for completion acceptance and technical pre-acceptance of completion.

In this code, Article 1.0.6 is compulsory provision, marked in bold type, and must be enforced strictly.

This code replaces the previous version SL 168—96.

This code is approved by the Ministry of Water Resources (MWR) of the People's Republic of China.

This code is interpreted by Bureau of Rural Hydropower and Electrification Development of MWR.

This code is chiefly drafted by National Research Institute for Rural Electrification of MWR.

This code is jointly drafted by Sichuan Local Electricity Bureau, and Zhejiang Hydropower Administration Center.

Chief drafters of this code are Liu Zhongmin, Fan Xinzhong, Lin Xuxin, Qiu Jianghai, Yu Zhenkai, Song Chao, Wu Jianzhang, Lv Yan, Lu Xiaoping, and Shu Jing.

This text of this code is published and distributed by China Water & Power Press.

The technical responsible person of this Code review meeting is Tang Tao.

The format examiner of this code is Cao Yang.

Contents

1 General Provisions

1.0.1 The main objectives of this code are to strength the management, ensure acceptance quality, systematize and standardize the acceptances in the small hydropower station construction.

1.0.2 This code is applicable to the acceptance of the construction of small hydropower stations with a total installed capacity ranging from 1.0 MW to 50 MW (hereinafter referred to as small hydropower project). It may be used as reference for the acceptance of the extension of small hydropower stations as well as the construction of hydropower stations under 1.0 MW.

1.0.3 According to the project plan and acceptance procedures, the acceptance of small hydropower projects can be classified into the acceptance of sub-unit projects, the acceptance of unit projects, the acceptance of contracted project completion, the stage acceptance (including the acceptance of turbine-generator startup), the special item acceptance and the completion acceptance. The acceptances shall be interconnected with each other and repetition shall be avoided.

According to the presiding organization, the acceptances of small hydropower projects can be classified as legal person acceptances and government acceptances. The legal person acceptances shall include the acceptance of sub-unit projects, the acceptance of unit projects, the acceptance of contracted project completion and the acceptance of the rest turbine-generator startup, while the government acceptances shall include the stage acceptance (including the acceptance of first and last turbine-generator startup), the special item acceptance and the completion acceptance. The presiding organization may add acceptance types and specific requirements if necessary.

1.0.4 The acceptances of small hydropower projects shall be based on the following items:

—Current laws, regulations, rules and technical standards of the country.

—Regulations of relevant departments.

—Approved project documents, design documents and corresponding revised design.

—Construction drawings, key equipment contracts and technical instructions.

—Construction contracts in case of legal person acceptances.

1.0.5 The acceptance of a small hydropower project shall include the following contents:

1 Examination of whether the completed part of the project meets the requirements of approved design documents.

2 Examination of the quality of the completed project regarding design, construction and equipment manufacture and installation as well as the situation regarding the collection, arrangement and filing of relevant materials.

3 Examination of whether the project meets the conditions for operation or further construction.

4 Examination of investment control and capital expenditure.

5 Suggestions for solving remaining problems of acceptance.

6 Comments for the project construction.

1.0.6 Acceptance of a project shall be organized in time when it meets all the requirements. If a project has not been accepted or

fails to pass acceptance, it shall not proceed with follow-up work or handover for commissioning.

1.0.7 A government acceptance shall be the responsibility of the acceptance committee established by the presiding organization, while a legal person acceptance shall be the responsibility of the acceptance working group set up by the legal person of the project. The acceptance committee (or working group) shall be formed by the representatives and experts from relevant organizations.

The acceptance appraisal report is the qualification document for acceptance, and shall be signed by all members of the acceptance committee (or working group). Any reserved comments on the conclusions shall be clearly recorded and signed in the appraisal report of acceptance.

1.0.8 Acceptance conclusions shall be approved by at least two thirds of all members of the acceptance committee (or working group).

Principles of solving problems found in the acceptance shall be consulted and ascertained by the acceptance committee (or working group). The chairman (or group leader) is authorized to give a final verdict on controversial issues. If over half of the members disagree with the verdict, it shall be referred to the supervision and management institution for a final decision in case of the legal person acceptance and shall be referred to the presiding organization for a decision in case of the government acceptance.

1.0.9 Concluding remarks shall be given specifically to the project quality after the inspection and assessment process.

1.0.10 Preparation of acceptance documentation shall be organized by the project legal person. Relevant organizations shall complete and submit them in time. The project legal person shall examine the completeness and normalization of the submitted acceptance documents.

1.0.11 Documents of acceptance are divided into those "shall be provided" and those "ready for check". The relevant organizations shall ensure the authenticity of the provided documents and bear the corresponding responsibility. Lists of documents needed to be provided or ready for check are shown in Annex A and Annex B respectively.

1.0.12 Drawings, data and final documents shall be prepared in line with the requirements for completion acceptance documents. Except drawings, the paper size should comply with international standard of A4 (210mm×297mm). Original copy of documents shall be stamped and shall not be replaced by any copies.

1.0.13 Acceptance fees shall be included in the project cost estimation and disbursed by the project legal person or paid according to the contract.

1.0.14 In addition to this code, the acceptance of a small hydropower project shall also conform to the current standards of the country.

2 Supervision and Management of Acceptance

2.0.1 The Ministry of Water Resources is in charge of the guidance on the supervision and management of the acceptances of nationwide small hydropower projects. According to prescribed limits of authority, the water administrative authorities of local people's governments at or above county level are responsible for supervision and management of acceptance of small hydropower project construction in their administrative areas.

2.0.2 According to the provisions of different levels of administration for local small hydropower projects, the water administrative authorities of local people's governments at all levels shall take charge of or participate in the government acceptances of small hydropower projects in their administrative areas, and undertake the supervision and management for the acceptance conducted by the legal person.

2.0.3 The approaches by which acceptances are supervised and managed shall include site inspection, organizing or participating in acceptance activities, filing acceptance work plan and output documents, etc.

If necessary, water administrative authorities and supervision and management organization for legal person acceptances shall take on-site inspections to the constructed projects and check the progress in the acceptance, and also make an investigation whenever any offence is reported.

2.0.4 The supervision and management for an acceptance shall include the following contents:

1 Whether the acceptance is in time.

2 Whether the acceptance meets the requirements.

3 Whether the composition of staff for acceptance meets the provisions.

4 Whether the acceptance procedure meets the specification.

5 Whether the acceptance documents are complete.

6 Whether the acceptance conclusion is clearly stated.

2.0.5 The authority being responsible for supervision and management shall make requirements to the presiding organization for a correction in a timely manner when an acceptance of project fails to meet the relevant rules and regulations. If necessary, the supervision and management authority may make a request to suspend the acceptance temporarily or redoing it and report to the presiding organization at the same time.

2.0.6 The authority being responsible for supervision and management for the acceptances conducted by the legal person shall check the received documents for the record, and ask the relative institutions to make correction, supplementation or completion if the document is deemed unqualified.

2.0.7 Project legal person shall prepare a working plan of acceptance 60 working days prior to the acceptance of first unit and report to the supervision and management authority for the record. If adjustment is made to the project construction plan, the working plan of acceptance shall also be updated and put on record again. The working plan shall include brief introduction of the project, project division, construction schedule and acceptance plan.

2.0.8 Technical problems found in the legal person acceptance shall in principle be settled according to the contract. If the provisions in the contract are not clear, the problems shall be settled according to the technical standards of the country. If no technical standards are available, the supervision and management authority for the legal person acceptance shall be responsible for a coordinated solution.

3 Acceptance of Sub-unit Projects

3.0.1 The acceptance of sub-unit projects shall be chaired by a legal person (or the supervisor entrusted by the legal person). The acceptance working group shall consist of the representatives of project legal person, designer, supervisor and contractor, and equipment manufacturers (suppliers).

3.0.2 The members of the acceptance working group shall have relevant professional knowledge or qualifications, and each institution should not nominate more than 2 representatives.

3.0.3 When a sub-unit project satisfies the acceptance conditions, the contractor shall submit an application of acceptance to the project legal person, including acceptance scope, inspection results of acceptance conditions and proposed time for acceptance. Project legal person shall reply within 5 working days after receiving the application and give an answer to the contractor for agreement or disagreement.

3.0.4 The acceptance of sub-unit projects shall meet the following conditions:

1 All item projects have been completed.

2 Construction quality of the completed item projects is all qualified. Relevant quality deficiencies have been dealt with or have obtained the treatment suggestions approved by the supervisor.

3 Other conditions written in the contract are satisfied.

3.0.5 The acceptance of sub-unit projects shall mainly include the followings:

1 Check whether the project meets the design standard or requirements of contract specification.

2 Evaluate the construction quality levels.

3 Make suggestions for problems discovered during the acceptance.

3.0.6 The acceptance of sub-unit projects shall follow the procedures below:

1 Listen to the report about construction and quality evaluation of item projects by the contractor.

2 Examine the completion situation of the project and construction quality on site.

3 Examine the quality evaluation of item projects and relevant files.

4 Discuss and pass the appraisal report (table) of acceptance of sub-unit projects.

3.0.7 The treatment situation for remaining problems in the acceptance of sub-unit projects shall have written records and be signed off by the representatives of relevant responsible institutions. Written records shall be filed together with the appraisal report of acceptance of sub-unit projects.

3.0.8 The format of appraisal report of acceptance of subdivided projects is shown in Annex C. The appraisal report of acceptance of subdivided projects shall be submitted to the relevant institutions by the legal person within 20 working days after its approval.

4 Acceptance of Unit Projects

4.0.1 Acceptance of unit projects shall be chaired by the project legal person (or the supervisor entrusted by the project legal person). The acceptance working group shall consist of representatives from project legal person, designer, supervisor, contractor, equipment manufacturers (suppliers) and operation and management institutions. If necessary, other experts may also be invited.

4.0.2 Members of the acceptance working group shall have relevant professional knowledge or occupational qualifications, and at least half of the members shall possess the professional title of engineer or senior engineer. Each institution should not nominate more than 2 representatives.

4.0.3 When the unit projects are completed and meet the acceptance conditions, the contractor shall submit an acceptance application to the legal person, with the contents as written in Article 3.0.3. The legal person shall determine whether to carry out the acceptance within 10 working days after receiving the application.

4.0.4 The legal person shall give advance notice to the quality and safety supervision organizations. For the unit project acceptance of key structures, advance notice shall be given to the supervision and management organizations for legal person acceptances. The supervision and management organizations for legal person acceptances may decide whether to send its representative to attend the acceptance meeting depending on specific situation, but the quality and safety supervision organization shall send its representative to the meeting.

4.0.5 The acceptance of unit projects shall meet the following conditions:

1 All sub-unit projects are completed and qualified.

2 The remaining problems found in the sub-unit project acceptances are solved and accepted, any remaining unsolved problems do not affect quality assessment of the unit projects and there are treatment suggestions approved by the supervisor.

3 Other terms in the contract are satisfied.

4.0.6 The acceptance of unit projects shall include:

1 Check whether the project has been completed in line with the approved design.

2 Assess quality of project construction.

3 Check the treatment situation for the remaining problems of the sub-unit project acceptance and relevant records.

4 Give treatment suggestions for problems of the acceptance.

4.0.7 The acceptance of unit projects shall follow the procedures below:

1 Listen to the construction reports given by the organizations involved in the project.

2 Examine the completion situation and project quality on site.

3 Examine relevant documents and files for the acceptance of the sub-unit projects.

4 Discuss and pass the appraisal report for the acceptance of unit projects.

4.0.8 Annex D is the format for unit projects acceptances. After the appraisal report of the acceptance of unit projects has been passed, it shall be sent within 20 working days to relevant organizations by the legal person and reported to the supervisor and management organization as well as quality and safety supervision organization for record keeping.

5 Acceptance of Contracted Project Completion

5.0.1 Acceptance of contract project completion shall be carried out after the construction works written in the contract have been completed. When the contracted project refer to only one unit project (or one sub-unit project), the unit project (or sub-unit project) acceptance should be carried out together with the acceptance of the contracted project completion, and shall meet relevant acceptance conditions.

5.0.2 Acceptance of contracted projects shall be chaired by a legal person. The acceptance working group shall consist of the legal person, representatives from designer, supervisor, contractor, operation and management organizations as well as major equipment manufacturers (suppliers). If necessary, other experts may be invited.

5.0.3 Members of the acceptance working group shall have relevant professional knowledge or occupational qualifications, and at least half of the members shall possess the professional title of engineer or senior engineer. Each institution should not nominate more than 2 representatives.

5.0.4 When the contracted project is completed and meets the acceptance conditions, the contractor shall submit an application to the legal person, with the contents as written in Article 3.0.3. The legal person shall decide whether to carry out the acceptance within 15 working days after the application report is received.

5.0.5 The acceptance of contracted project completion shall meet the following conditions:

1 All projects and work within the scope of contract have been completed.

2 Projects have been accepted and qualified according to stipulations.

3 Initial values and all monitoring data during construction period have been recorded by instruments and equipment.

4 Quality defects have been corrected according to the requirements and accepted.

5 Settlement account of project completion has been closed.

6 Construction site has been cleared.

7 Files that shall be transferred to the legal person have been sorted out.

8 Other conditions written in the contract are satisfied.

5.0.6 The acceptance of contracted project completion includes the following contents:

1 Examination of the completion situation of the project and works in the contract.

2 Examination of the clearing condition of the construction site.

3 Examination of the operation situation for commissioned projects.

4 Examination of the collection situation for acceptance data and documents.

5 Assessment of quality levels of the construction.

6 Examination of the completion settlement account of the project.

7 Examination of the remaining problems and the treatment situations for previous acceptances.

8 Treatment suggestions for problems found in the acceptance.

9 Determination of the completion date of the contracted project.

10 Discussion and passing of the appraisal report of completion acceptance of the contracted project.

5.0.7 Annex E shows the appraisal report format for the acceptance of contracted project completion. After the appraisal report of the acceptance of contract project completion has been passed, it shall be sent within 20 working days to relevant organizations by the legal person and reported to the acceptance supervision and management organization as well as the quality and safety supervision organization for record keeping.

6 Stage Acceptance

6.1 General Requirements

6.1.1 The stage acceptance shall include diversion (closure) acceptance, impounding acceptance, acceptance of turbine-generator startup and others requested by the presiding organisation according to the needs of the construction project.

6.1.2 The stage acceptance shall be chaired by the presiding organization or its entrusted organization for completion acceptance. The stage acceptance committee consists of representatives and relevant experts from the acceptance presiding organization, quality and safety supervision organization and operation and management organization. Officials from the local government and relevant departments may be invited if it is necessary.

The organizations involved in the project shall appoint representatives to attend the stage acceptance and sign on the appraisal report as the organizations being accepted.

6.1.3 When the construction is ready for the stage acceptance, the legal person shall submit an application (see Annex F for contents). The application shall be reviewed by the supervision and management organization and then reported to presiding organization who should give a decision for an agreement or a disagreement within 15 working days after it is received.

6.1.4 Stage acceptance shall include the following contents:

1 Examination of the appearance and quality of completed projects.

2 Examination of the construction situation.

3 Examination of uncompleted projects and implementation of main technologies as well as the construction conditions.

4 Examination of operation conditions for projects planned to be commissioned.

5 Examination of remaining problems and treatment situation of the previous acceptances.

6 Assessment of the quality of completed works.

7 Tackling suggestions for problems found in the acceptance.

8 Discussion and passing of the appraisal report for stage acceptance.

6.1.5 Stage acceptance shall include the following procedures:

1 An on-site investigation should be conducted on the construction situation and relevant data should be checked and read.

2 Meetings should be held for making:

1) Announce the list of stage acceptance committee members.
2) Listen to the work reports by the organizations involved in the project.
3) Discuss and pass the appraisal report for stage acceptance.
4) Ask representatives from the acceptance committee and organizations to sign the appraisal report.

6.1.6 Annex G shows the appraisal report format for stage acceptances. The appraisal report shall be sent to relevant institutions by the presiding organizations, within 20 working days after getting its approval.

6.2 Diversion (Closure) Acceptance

6.2.1 Before diversion (closure), the diversion (closure) acceptance shall be carried out. The diversion (closure) acceptance may be chaired by the presiding organization of the completion acceptance or by the entrusted legal person according to the size and importance of the project.

6.2.2 The diversion (closure) acceptance shall satisfy the following conditions:

1 Diversion works has been completed and ready for passing through water. It does not affect other construction when it is in use.

2 Underwater concealed works for diversion is completed and has passed acceptance.

3 River closure plan is done and all preparations are finished.

4 Construction plan for flood period has been approved by the authorized flood control headquarter and relevant measures have been adopted.

5 Relocation of resettlement people and clear-up of the reservoir bottom under raised water level after river closure have been completed and passed acceptance.

6 Obstruction and barriers on navigable rivers are removed.

6.2.3 The diversion (closure) acceptance shall include the followings:

1 Examine whether underwater works, concealed works and diversion works (river closure works) meet the requirements for the diversion (closure).

2 Examine the completion situation of land requisition, resettlement and clearing up of the reservoir bottom.

3 Review the closure plan, the measures and the preparations for the diversion (closure).

4 Examine the implementation of measures for tackling

navigation obstruction problems.

5 Assess the quality of completed works related to the river closure.

6 Give tackling suggestions for problems found in acceptance.

7 Discuss and pass the appraisal report for stage acceptance.

6.2.4 If the diversion works (river closure) is divided into several phases, the diversion (closure) acceptance should also be carried out in stages.

6.3 Acceptance of Reservoir Impoundment

6.3.1 The acceptance of reservoir impoundment shall be carried out before the impoundment. The acceptance of reservoir impoundment shall be chaired by the presiding organization of completion acceptance or the entrusted legal person according to the scale and importance of the project.

6.3.2 The acceptance of reservoir impoundment shall meet the following conditions:

1 The external appearance of water retaining structures meets the requirements of the impounding water level.

2 Resettlement of affected residents and reservoir bottom clearing has been completed and pass the acceptance.

3 The discharge structure used after impoundment is almost completed and satisfies requirements of overflowing.

4 Instruments and equipment for observation have been installed and tested according to the design, and the initial values and observation data during construction period have been obtained.

5 Construction plan and measures for uncompleted works

after impoundment have been worked out.

6 The appraisal report of impounding safety has been submitted in line with the regulations and a specific conclusion has been reached to allow the impoundment in the report.

7 Issues that may affect the safe operation after impoundment have been solved and solutions for tackling key technical problems have been addressed.

8 The impoundment plan and the diversion outlet (tunnel) plugging plan have been compiled and approved, and all preparations are complete.

9 Annual construction safety scheme during flood season (including flood regulating plan) has been approved by the authorized flood control headquarters and related measures have been carried out.

6.3.3 The acceptance of reservoir impoundment shall include the followings:

1 Check whether the completed works meets impounding requirements.

2 Check land requisition, resettlement of affected residents and reservoir clearing.

3 Check the treatment condition of surrounding banks near the dam.

4 Check impounding preparation.

5 Appraise the construction quality of completed construction related to impoundment.

6 Give tackling suggestions for problems of the acceptance.

7 Discuss and pass the appraisal report of stage acceptance.

6.3.4 If impoundment is carried out in stages, the acceptance of reservoir impoundment should also be carried out in stages.

6.4 Acceptance of Turbine-generator Startup

6.4.1 The turbine-generator startup acceptance shall be carried out before each turbine-generator unit is put into operation.

6.4.2 The acceptance of turbine-generator startup shall satisfy the following conditions:

1 Hydraulic structures related to turbine-generator startup are completed and meet the requirements of turbine-generator startup.

2 The water level in reservoir (headrace) is higher than the lowest water level for power generation and the water quantity to be diverted can satisfy the minimum requirements for turbine-generator startup.

3 Metal structures and hoisting equipment related to turbine-generator startup are installed and tested to meet the requirements of turbine-generator startup.

4 The turbine-generator unit, auxiliary equipment and other oil, gas and water accessories are installed and tested. Each part has been put into commissioning and all of which can meet the requirements of turbine-generator startup.

5 Relevant electrical equipment (or devices) are installed and tested according to relevant norms to meet the requirements of turbine-generator startup.

6 Electrical transmission, transformer equipment and facilities are constructed, installed and tested, and have passed safety evaluation or acceptance by relevant departments. Preparatory work for power transmission has been carried out and meets the requirements of turbine-generator startup.

7 Measuring, monitoring, control and protection of electrical equipment for turbine-generator startup are installed and tested.

8 Operation management organization is set up and the staffs are employed and meet the requirements for turbine-generator startup.

9 Protective measures related to safety and fire control for turbine-generator startup are put into practice.

10 Rules and regulations related to on-site safety operation specifications are drawn up.

6.4.3 The acceptance of the first (last) turbine-generator startup shall be chaired by the turbine-generator startup acceptance committee established by the presiding organization of completion acceptance or by an organization entrusted. The acceptance for the rest turbine-generators' startup shall be chaired by the turbine-generator startup acceptance group organized by the legal person. Representatives of local grid enterprises shall also join the acceptance committee (group).

6.4.4 Under the turbine-generator startup committee, a commissioning commanding group and an acceptance handover group shall be set up for undertaking specific work.

The project technical leader of the installation contractor serves as the leader of the commissioning commanding group while the technical leader of the operation and management organization as the vice group leader. The group shall be responsible for compilation of turbine-generator unit commissioning test documents, and organizing commissioning and examination. Operators on duty for turbine-generator unit commissioning shall consist of staff from the unit installation contractor, operation and management organization and major equipment manufacturers (suppliers).

The legal person serves as the leader of the acceptance handover group while the operation management organization, contractor and supervisor as the vice group leaders. The group shall consist of staff from the operation management organization, contractors for civil works and equipment installation, major equipment manufacturers (suppliers) and supervisor. The group is responsible for the inspection on completion situation and quality of civil works, metal structures and equipment installation as well as handovers of technical documents, drawings, spare parts and special tools etc.

6.4.5 The acceptance committee for turbine-generator startup has the following major work items:

1 Listen to the reports of project legal person, designer, supervisor, contractor and operation management organization as well as the commissioning command group and acceptance handover group; review related documents; examine whether the status of the installation of turbine-generator, accessory equipment, electrical equipment and construction of civil works and quality meet the design requirements and standards stipulated by the contracts, and whether they satisfy the requirements of turbine-generator startup.

2 Check all preparations before turbine-generator startup and the requirements in Article 6.4.2 and the others proposed by the acceptance committee. Tackling suggestions for unqualified items and remaining problems shall be proposed.

3 Review and approve the turbine-generator startup test procedure, operation code and commissioning plan; determine the date of the first time turbine-generator startup.

4 Prepare appraisal report of startup acceptance, and determine the list of handover items.

6.4.6 For commissioning of turbine-generator startup, a turbine-generator unit startup test and test of the unit with rated load running for 72 hours continuously shall be carried out.

1 Turbine-generator unit startup tests. The procedure of the startup test shall be compiled by the commissioning command group and approved by the startup acceptance committee, and includes:

1) Examine and test the diversion system, turbine and governor, generator and excitation system, oil-water-gas system and generator cooling system, electrical equipment, control protection devices, and measuring meters.

2) Examine and test diversion facilities and equipment both during and after water filling.

3) Examine and test the first time startup of the turbine-generator unit and no-load operation.

4) Examine and test the turbine-generator unit when connecting to the grid and operation with loads.

5) Test turbine-generator unit load rejection.

2 Tests of the turbine-generator unit with rated load running for 72 hours continuously. Before the test, the committee shall listen to a brief report by the commissioning command group and the supervisor, so as to make a decision on whether the unit can run 72 hours with rated load continuously. If the actual load less than the rated output of the unit or the output less than the rated one due to some special reasons, the committee may determine the maximum test load of the unit according to actual conditions.

3 If the unit running for 72 hours with loads is normal, then turbine-generator startup commissioning is deemed completed. The commissioning command group shall report the completion situation of commissioning to the committee and prepare a working report of the turbine-generator startup commissioning.

6.4.7 During the turbine-generator startup commissioning, records for unit examination, test and commissioning activities shall be kept. All of these records shall form part of the technical documents to be submitted to the operation and management organization.

6.4.8 Equipment defects and faults found in the commissioning shall be solved by the responsible organization in time. Those unqualified items cannot be handed over for trial commercial operation.

6.4.9 When the turbine-generator unit can run safely for trial commercial operation confirmed through its startup commissioning, the committee shall issue the appraisal report of acceptance of turbine-generator startup (see the format in Annex H).

6.4.10 When the appraisal report of acceptance of turbine-generator startup is issued, unit handover from the contractor to the operation and management organization shall be carried out for trial commercial operation. The trial commercial operation period ranges from 6 months (one flood period covered) to 12 months.

6.4.11 Acceptance of the rest turbine-generator startup may refer to the requirements of acceptance of first and last turbine-generator startup as reference, and shall be carried out by the commissioning command group and acceptance handover group organized by the legal person. If a failure or weakness found during the process of acceptance, it shall be reported to the presiding organization of completion acceptance in a timely manner.

7 Acceptances of Special Items

7.0.1 Before the completion acceptance, the acceptance of special items shall be carried out based on relevant national and local rules. The presiding organizations of special item acceptances shall be determined according to the rules of the country or its relevant departments.

7.0.2 Project legal person shall submit an application for the special item acceptance to the relevant department and make necessary preparation in accordance with the rules of the country and its relevant departments in charge.

7.0.3 Necessary requisites of special item acceptance, acceptance contents, acceptance procedures and requirements for output documents shall conform to the provisions of the country and its relevant departments.

8 Completion Acceptance

8.1 General Requirements

8.1.1 When a constructed project satisfies the conditions of completion acceptance, the legal person shall submit an application report to the legal person acceptance supervision organization and presiding organization (see Annex I for more details).

8.1.2 The completion acceptance shall meet the following conditions:

1 The project has been completed according to the approved design.

2 Major design changes have been approved by the organization with authority.

3 Each of the unit projects can be operated normally and all the turbine-generators are in operation.

4 Turbine-generator commissioning term has ended, and hydraulic structures have experienced one flood season and one freezing season.

5 Problems discovered in the previous acceptances have been solved.

6 Acceptances of all special items have been approved.

7 Working report on quality and safety supervision has been submitted, and project quality is qualified.

8 Financial reports of final accounts of state-financed projects have been audited, and the problems found by the audit have been fixed and the report of the fixed problems has been submitted.

9 Documents for completion acceptance are ready for inspection, and the format for the completion acceptance working report and its main contents are shown in Annex K and Annex L.

8.1.3 If the completion acceptance of the project fails to proceed on schedule, the legal person shall submit a special application report to the presiding organization for delayed acceptance. The application report shall include the main reasons of delay acceptance and expected extension of time.

8.1.4 If there is still an amount of uncompleted work, which does not affect normal operation and meets relevant financial provisions, and the legal person has made necessary arrangements for the uncompleted work with the approval of the presiding organization, the completion acceptance may be carried out.

8.1.5 The completion acceptance is divided into two stages including the technical pre-acceptance of completion and the completion acceptance.

8.1.6 The completion acceptance shall follow the procedures below:

1 The legal person organizes self-inspection for the completion acceptance.

2 The legal person submits the application report of completion acceptance.

3 The presiding organization approves the application report of completion acceptance.

4 The technical pre-acceptance of completion is carried out.

5 A completion acceptance meeting is held.

6 The appraisal report of completion acceptance is printed and dispatched.

8.2 Self-inspection for Completion Acceptance

8.2.1 The legal person shall organize self-inspection before applying for the completion acceptance. The self-inspection shall be chaired by the legal person, and with the representatives from designer, supervisor, contractor and equipment manufacturers (suppliers) as well as the operation management organization.

8.2.2 The self-inspection for completion acceptance includes:

1 Check working reports of the relevant organizations.

2 Examine the construction situation and evaluate the quality levels of the project.

3 Check the remaining problems found in the previous special item acceptances and treatment of problems during the trial commercial operation.

4 Determine the contents and term of unfinished work and its accordingly responsible organization.

5 Make arrangements for the work to be done before the completion acceptance.

6 Discuss and pass the self-inspection working report of completion acceptance.

8.2.3 10 working days before the self-inspection for completion acceptance, the legal person shall give advance notice to the quality and safety supervision organization as well as report to the supervision and management organization. The representatives from quality and safety supervision organization shall attend the self-inspection working meeting.

8.2.4 Project legal person shall report the quality conclusion and relevant documents of self-inspection to the quality supervision organization within 10 working days after the self-inspection for completion acceptance is completed.

8.2.5 Format of self-inspection working report is shown in Annex J. People who attend the self-inspection shall sign their names on the working report. The legal person shall report to the supervision and management organization within 20 working days after the self-inspection working report passed.

8.3 Inspection of Project Quality by Sampling

8.3.1 According to the requirements for the completion acceptance, the presiding organization may entrust a qualified quality inspection organization to carry out the inspection of project quality by sampling. The legal person shall sign the project quality inspection agreement with the quality inspection organization. Inspection fees shall be borne by the legal person, but the inspection fees for unqualified items shall be paid by the responsible organization.

8.3.2 The project quality inspection organization shall not be the same entity as the legal person, designer, supervisor, contractor or manufacturers (suppliers).

8.3.3 Based on the requirements of the presiding organization and the situation of the project, the project legal person shall proposes the items, contents and the number of the inspection for sampling, and then report to the project quality supervision organization for examination and finally to the presiding organization for verification.

8.3.4 The project quality inspection organization shall inspect the project quality based on relevant technical standards, and provide the quality inspection report in time according to the contract. The legal person shall submit the inspection report to the presiding organization within 10 working days after receiving the inspection report.

8.3.5 The legal person shall organize relevant organizations to solve the problems discovered by sampling tests in time. The completion acceptance shall not proceed if the problems that affect safe operation are not solved.

8.4 Technical Pre-acceptance of Completion

8.4.1 Technical pre-acceptance of completion shall be chaired by an expert panel of the presiding organization. More than two-thirds of the panel members shall be engineer or relevant professional qualifications, and over one-third shall be senior engineer or relevant professional qualifications. More than two-thirds of the panel members shall come from non-project-involved organizations. Representatives from the project-involved organizations shall attend the technical pre-acceptance and answer questions from the expert panel.

8.4.2 Under the expert panel of the technical pre-acceptance of completion, professional working groups may be set up. The pre-acceptance working report will be worked out on the basis of the comments from each working group.

8.4.3 The technical pre-acceptance of completion shall include the following contents:

1 Check whether the project is completed based on the approved design.

2 Examine whether the project has potential quality defects and problems that affect safe operation.

3 Examine the remaining problems discovered in the previous special acceptances and the treatment situation for problems found in trial commercial operation.

4 Evaluate the key technical problems of the project.

5 Examine arrangements for unfinished works.

6 Appraise the project quality.

7 Examine the investment and financial situation of the project.

8 Propose the solutions to the problems discovered in the acceptance.

8.4.4 The technical pre-acceptance of completion shall follow the procedures below:

1 Examine the construction situation on site and review relevant construction documentation.

2 Listen to working reports from the legal person, the designer, supervisor and contractor, the quality and safety supervision organization and the operation management organization.

3 Listen to the sampling inspection report for project quality.

4 Professional working groups to discuss and give suggestions.

5 Discuss and pass the technical pre-acceptance working report for completion.

6 Draft the appraisal report of completion acceptance.

8.4.5 Working report of technical pre-acceptance for completion shall be an attachment of the appraisal report of completion acceptance, and the format is shown in Annex M.

8.5 Completion Acceptance

8.5.1 The completion acceptance committee shall consist of the representatives and experts of the presiding organization, local people's government, water administrative department, quality and safety supervision organization, investors and operation management organization. The committee may have chair person and several deputies and members. The chair person shall be one of the representatives from the presiding organization.

8.5.2 The legal person, designer, supervisor, contractor and the equipment manufacturers (suppliers) shall have representatives to participate in the completion acceptance, answer questions from the committee and sign on the appraisal report of acceptance.

8.5.3 The completion acceptance meeting shall include the following procedures:

—Check construction situation on site and review relevant documents.

—The meeting shall be held to:

1) Announce the name list of the acceptance committee.

2) Listen to the working report of the construction management.

3) Listen to the working report on the technical pre-acceptance of completion.

4) Listen to other reports designated by the acceptance committee.

5) Discuss and pass the appraisal report of the completion acceptance.

6) Ask members and representatives from the acceptance committee and acceptance organization to sign on the appraisal report of the completion acceptance.

8.5.4 If project quality reaches the qualified level or above, the quality conclusion of the completion acceptance shall also be qualified.

8.5.5 The appraisal report format for the completion acceptance is shown in Annex N. The appraisal report of the completion acceptance shall be submitted to the relevant organizations by the presiding organization within 20 working days after it is approved.

9 Project Handover and Handling of Leftover Problems

9.1 Project Transfer and Handover

9.1.1 When the completion acceptance of a contract project has been approved, the legal person and contractor shall nominate persons responsible for project transfer and handover within 20 working days, with complete written records and signatures of both parties.

9.1.2 The legal person and the contractor shall finish transfer and handover of the project and file documents within specified time in the construction contract or the appraisal report of acceptance.

9.1.3 During the transfer and handover of specific project, the contractor shall submit warranty for project quality to the legal person and the format of warranty is shown in Annex O. The contents of warranty shall conform to the terms and reference of the contract.

9.1.4 The validity of warranty shall commence from the completion acceptance, except for those contracts which stipulate otherwise.

9.1.5 Project legal person shall issue a certificate of project completion to the contractor within 20 working days after the warranty of project quality and other documents related to project completion have been provided and clear up of construction site has been finished. The format of the certificate is shown in Annex P.

9.1.6 After handover, the legal person shall transfer the project to the operation management organization in time. Project transfer shall include the project entity, other fixed assets and file and documents of the project, which shall be checked one by one according to relevant approved documents such as the preliminary design and go through transfer formality, and the entire process of transfer shall be recorded in written form.

9.2 Treatment of Remaining Problems with Acceptance and Uncompleted Work

9.2.1 Remaining issues after project acceptance shall be clearly stated in the relevant output documents. Those that affect normal operation of the project shall not be treated as remaining issues.

9.2.2 The treatment of remaining problems with acceptance and uncompleted works shall be the responsibility of the legal person. The legal person shall supervise the relevant responsible organizations to solve all problems according to the requirements of the appraisal report and contract.

9.2.3 After the remaining problems with acceptance and uncompleted works construction are finished, the legal person shall organize acceptance, work out the output documents and report to the presiding organization.

9.2.4 After the completion acceptance, the legal person shall be responsible for the treatment of remaining problems. If the legal person is repealed, the investor or organization that set up the legal person or its appointed organization is responsible for treatment.

9.3 Issuance of Completion Certificate

9.3.1 If the contractor completes the treatment of quality defects in the scope of its warranty within the quality warranty period, the legal person shall issue a warranty termination certificate of project quality to the contractor within 20 working days after the expiration of warranty. The format is shown in Annex Q.

9.3.2 After the expiration of quality warranty period and the completion of the remaining problems and the uncompleted works, the legal person shall make an application to the presiding organization for the completion certificate, with the following contents:

1 Information about project transfer and handover.

2 Information about project operation and management.

3 Treatment of the remaining problems and construction of the uncompleted works.

4 Relevant warranty situation for project quality.

9.3.3 The presiding organization for completion acceptance shall determine whether to issue the completion certificate within 20 working days after the legal person application report is received. The completion certificate format is shown in Annex R (Original) and Annex S (Copy). Following conditions should be met when the certificate is issued:

1 Appraisal report of completion acceptance has been issued.

2 Remaining issues have been solved and unfinished work has been done and passed the acceptance.

3 The entire project has been delivered to the operation management organization.

9.3.4 The completion certificate shall be issued with 3 originals and several copies, of which, the original ones shall be kept by the legal person, the operation management organization and the file department respectively, and the copies shall be kept by the major participating organizations.

Annex A List of Required Documents for Acceptances

Table A List of required documents for acceptances

No.	Item	Acceptance of sub-unit projects	Acceptance of unit projects	Acceptance of contracted project completion	Acceptance of turbine-generator startup	Stage acceptance	Technical pre-acceptance	Completion acceptance	Organization
1	Working report for construction management			√	√	*	√	√	Project legal person
2	Construction chronicle			*	√	*	√	√	Project legal person
3	List of projects for acceptance	√	√	√	√	√	√	√	Project legal person
4	List of unfinished works, construction arrangement and planned completion time			√	√	√	√	√	Project legal person
5	Flood safety plan in flood season				*	√	√	√	Project legal person
6	Operation plan for project scheduling					√	√	√	Project legal person
7	Special report for major technical problems					*	*	*	Project legal person
8	Appraisal report for acceptance (draft)				√	√	√	√	Presiding organization

Table A (Continue)

No.	Item	Acceptance of sub-unit projects	Acceptance of unit projects	Acceptance of contracted project completion	Acceptance of turbine-generator startup	Stage acceptance	Technical pre-acceptance	Completion acceptance	Organization
9	Working report for construction supervision			√	√	√	√	√	Supervisor
10	Working report for design of the project			*	√	*	√	√	Designer
11	Working report for construction management			√	√	√	√	√	Contractor
12	Plan for turbine-generator startup and commissioning				√				Contractor
13	Working report for turbine-generator commissioning				√				Contractor
14	Report for project quality and safety supervision				√	*	√	√	Quality and safety supervision organization
15	Working report for operation management						√	√	Operation management organization
16	Working report for technical pre-acceptance						√	√	Experts panel
17	Technical appraisal report for completion acceptance						*	*	Technical appraisal organization

Note: "√" refers to "shall be provided"; "*" refers to "should be provided" or "be provided according to demand".

Annex B List of Documents for Reference in Acceptances

Table B List of documents for reference in acceptances

No.	Item	Acceptance of sub-unit project	Acceptance of unit project	Acceptance of contracted project completion	Acceptance of turbine-generator startup	Stage acceptance	Technical pre-acceptance	Completion acceptance	Organization
1	Preliminary work and official written documents		√	√	√	√	√	√	Project legal person
2	Competent department remarks		√	√	√	√	√	√	Project legal person
3	Tendering and bidding documents		√	√	√	√	√	√	Project legal person
4	Contract documents		√	√	√	√	√	√	Project legal person
5	Project classification documents	√	√	√	√	√	√	√	Project legal person
6	Quality evaluation documents of sub-unit projects		√	*	√	√	√	√	Project legal person
7	Quality evaluation documents of unit projects		√	*			√	√	Project legal person
8	Quality evaluation documents of project appearance		√				√	√	Project legal person

Table B (Continue)

No.	Item	Acceptance of sub-unit project	Acceptance of unit project	Acceptance of contracted project completion	Acceptance of turbine-generator startup	Stage acceptance	Technical pre-acceptance	Completion acceptance	Organization
9	Major minutes	√	√	√	√	√	√	√	Project legal person
10	Safety and quality accident documents	√	√	√	√	√	√	√	Project legal person
11	Appraisal reports for stage acceptances						√	√	Project legal person
12	Final completion accounts and audit documents						√	√	Project legal person
13	Relevant documents for special item acceptances						√	√	Project legal person
14	Technical appraisal report for project safety					√	√	√	Project legal person
15	Documents of detail design		√	√	√	√	√	√	Designer
16	Design changes of the project	√	√	√	√	√	√	√	Designer
17	Supervision documents	√	√	√	√	√	√	√	Supervisor
18	Record tables of quality defects	√	√	√	√	√	√	√	Supervisor
19	Quality evaluation documents of item projects	√	√	√	√	√	√	√	Contractor

Table B (Continue)

No.	Item	Acceptance of sub-unit project	Acceptance of unit project	Acceptance of contracted project completion	Acceptance of turbine-generator startup	Stage acceptance	Technical pre-acceptance	Completion acceptance	Organization
20	Quality inspection documents of project construction	√	√	√	√	√	√	√	Contractor
21	Built drawings		√	√		√	√	√	Contractor
22	Documents of land requisition and resettlement		√			√	√	√	Sponsor
23	Documents for project quality management	√	√	√	√	√	√	√	Participating organizations
24	Documents for project safety management	√	√	√	√	√	√	√	Participating organizations
25	Technical standards used for the project	√	√	√	√	√	√	√	Participating organizations
26	Mandatory articles of standards	√	√	√	√	√	√	√	Participating organizations
27	Others	To be provided by relevant organizations according to demand							

Note: "√" refers to "shall be provided"; "*" refers to "should be provided" or "be provided according to demand".

Annex C Format of Appraisal Report for Acceptance of Sub-unit Project

No. :

××× Hydropower Project

Acceptance of ××× Sub-unit Project

Appraisal Report

Name of the unit project:

Working group of acceptance for ××× sub-unit project

Date

Foreword (including acceptance basis, organizations and process)

1 Commencement and completion dates of the sub-unit project

2 Construction contents of the sub-unit project

3 Construction process and major finished quantities

4 Quality accidents and treatment for quality defects

5 Quality evaluation of projects for acceptance (including item projects, the number of major item projects, rate of "qualified" and "excellent"; self-evaluation results of the contractor; re-evaluation results of the supervisor; quality level for sub-unit project)

6 Remaining problems discovered in the acceptance and their treatment comments

7 Conclusions

8 Reservation (signature of the person with reservations)

9 Signature list of working group members for sub-unit project acceptance

10 Annex: Treatment records for remaining problems discovered in the acceptance

Annex D Format of Appraisal Report for Acceptance of Unit Project

××× Hydropower Project

Acceptance of ××× Unit Project

Appraisal Report

Working group of acceptance for ××× unit project

Date

Presiding organization for acceptance:

Supervision and management organization of legal person acceptance:

Legal person:

Construction agent (if any):

Designer:

Supervisor:

Contractor:

Major equipment manufacturer (supplier):

Quality and safety supervision organization:

Operation and management organization:

Acceptance date:

Acceptance address:

Foreword (including acceptance basis, organizations and process)

1 Unit project overview

1.1 Name and location of the unit project

1.2 Major construction contents of the unit project

1.3 Construction process of the unit project (including commencement and completion dates, major measures adopted in construction)

2 Acceptance scope

3 Completion situation and major finished quantities

4 Quality evaluation of the unit project

4.1 Quality evaluation of the sub-unit project

4.2 Quality evaluation of the project appearance

4.3 Project quality test situation

4.4 Quality level for the unit project

5 Treatment of remaining problems discovered in the acceptance of the sub-unit project

6 Operation preparation situation (in case of acceptance of project commission)

7 Major problems and their solution comments

8 Comments and suggestions

9 Conclusion

10 Reservation (signature of the person with reservations)

11 Signature list of members of work group for the unit project acceptance

Annex E Format of Appraisal Report for Acceptance of Contracted Project Completion

××× Hydropower Project

Acceptance of ××× Contracted Project Completion

(Contract title and serial number)

Appraisal Report

Working group of acceptance for ××× contracted project completion

Date

Legal person:

Construction agent (if any):

Designer:

Supervisor:

Contractor:

Major equipment manufacturer (supplier):

Quality and safety supervision organization:

Operation and management organization:

Acceptance date:

Acceptance address:

Foreword (including acceptance basis, organizations and process)

1 Contracted project overview

1.1 Name and location of the contracted project

1.2 Major construction contents of the contracted project

1.3 Construction process of the contracted project (including commencement and completion dates, major measures adopted in construction)

2 Scope of acceptance

3 Contract implement situation (including contract management, project completion situation and major finished quantities, accounts settlement situation, etc.)

4 Treatment of remaining problems discovered in the previous acceptances

5 Quality evaluation of the contracted project

6 Major problems and treatment suggestions

7 Comments and suggestions

8 Conclusions

9 Reservations (signature of the person with reservations)

10 Signature list of members of working group for the contracted project acceptance

11 Annex: List of documents transferred by the contractors to the legal person

Annex F Requirements for Application for Stage Acceptances

1 General situation of the project

2 Examination results for project acceptance conditions

3 Preparation situation for project acceptance

4 Suggested acceptance date, location and participating organizations

Annex G Format of Appraisal Report for Stage Acceptances

××× Hydropower Project

Acceptance of ××× Stage

Appraisal Report

Committee (working group) for ××× stage acceptance of ××× hydropower project

Date

Presiding organization for acceptance:

Supervision and management organization of legal person acceptance:

Legal person:

Construction agent (if any):

Designer:

Supervisor:

Major contractors:

Major equipment manufacturer (supplier):

Quality and safety supervision organization:

Operation and management organization:

Acceptance date:

Acceptance address:

Foreword (including acceptance basis, organizations and process)

1 Project overview

1.1 Project location and major development purposes

1.2 Major technical indices of the project

1.3 Project design overview (including design approval situation, major design quantities and investment)

1.4 Construction overview (including construction and completed quantities)

2 Acceptance scope and contents

3 Project appearance (corresponding to the acceptance scope and completion situation)

4 Quality evaluation of the project

5 Completed works (including safety appraisal, resettlement of affected residents and acceptance of reservoir bottom cleaning, technical pre-acceptance)

6 Arrangement for river closure (reservoir impounding)

7 Plan for flood protection and dispatching

8 Construction plan for the unfinished project

9 Major problems and treatment comments

10 Suggestions

11 Conclusions

12 Signature list of members of committee (work group)

13 Annex: Working report for technical pre-acceptance (if any)

Annex H Format of Appraisal Report for Acceptance of Turbine-generator Startup

××× Hydropower Project

Acceptance of Turbine-generator Startup

Appraisal Report

Committee (working group) for turbine-generator startup acceptance of ××× hydropower project

Date

Presiding organization for acceptance:

Supervision and management organization of legal person acceptance:

Legal person:

Construction agent (if any):

Designer:

Supervisor:

Major contractor:

Major equipment manufacturer (supplier)

Quality and safety supervision organization:

Operation and management organization:

Acceptance date:

Acceptance address:

Foreword (including acceptance basis, organizations and process)

1 Project overview

1.1 Major construction contents of the project

1.2 Major technical indices of the unit

1.3 Design, manufacture and installation of the turbine-generator and auxiliary equipment

1.4 Project appearance related to turbine-generator startup

2 Scope and contents of acceptance

3 Quality evaluation for the project

4 Commissioning situation of the turbine-generator startup

5 Treatment suggestions for project defects and remaining problems

6 Notices and suggestions for project handover

7 Conclusions

8 Signature list of members of committee (working group)

9 Annex: Working report of turbine-generator commissioning

Annex I Requirements for Application for Project Completion Acceptance

1 Project general situation

2 Examination results for project acceptance conditions

3 Uncompleted work and construction plan

4 Preparations for project acceptance

5 Proposed acceptance date, location and participating organizations

6 Annex: Self-inspection report for project completion acceptance

Annex J Format of Self-inspection Report for Project Completion Acceptance

Completion Acceptance of ××× Hydropower Project

Self-inspection Report

Self-inspection working group for completion acceptance of ××× hydropower project

Date

Legal person:

Construction agent (if any):

Designer:

Supervisor:

Major contractor:

Major equipment manufacturer (supplier):

Quality and safety supervision organization:

Operation and management organization:

Foreword (including acceptance basis, organizations and process)

1 Project overview

1.1 Name and location

1.2 Major construction contents

1.3 Construction process

2 Completion situation of the project

2.1 Completion situation

2.2 Comparison between completed and approval quantities

2.3 Finished acceptances

2.4 Investment and audit

2.5 Handover and commissioning of project

3 Quality evaluation of the project

4 Treatment of remaining problems with the acceptance

5 Uncompleted works and construction plan

6 Major problems and suggestions

7 Conclusions

8 Signature list of members of self-inspection working group for project completion acceptance

Annex K Format of Working Report for Project Completion Acceptance

Completion Acceptance of ×××
Hydropower Project

××× Working Report

Prepared by:

Date

Approval：

Review：

Examination & verification：

Major compilers：

Annex L Format of Working Report Content for Project Completion Acceptance

L. 1 Working Report for Project Management

L. 1. 1 Project overview

1 Project location

2 Project proposal and approval documents of preliminary design

3 Project purposes and design criteria

4 Project characteristics

5 Major construction contents of the project

6 Project layout

7 Project investment

8 Major quantities and total period of construction

L. 1. 2 General of project construction

1 Construction preparation

2 Tendering and Participating organizations

3 Commencement report and approval reply for the project

4 Commencement and completion dates of major projects

5 Construction process of major structure

6 Major design changes

7 Treatment of major technical problems

8 Flood control during construction

L. 1. 3 Special items of project

1 Land requisition compensation and resettlement of affected residents

2 Environmental protection projects

3 Water and land conservation facilities

4 Construction files

L. 1. 4 Project management

1 Organization establishment and work situation

2 Process of bidding for major projects

3 Project budget and completion situation of investment plan

1) Approved budget and real performance

2) Annual plan and performance

3) Investment source and fund situation

4 Contract management

5 Materials and equipment supply

6 Capital management and final payment

L. 1. 5 Project quality

1 Quality management and supervision of the project

2 Project classification

3 Quality control and check

4 Treatment situation for quality accidents

5 Quality level evaluation

L. 1. 6 Safety in production and civilized construction site

L. 1. 7 Project acceptances

1 Acceptance of unit projects

2 Stage acceptances

3 Special item acceptances

L. 1. 8 Appraisal for reservoir impounding safety, and technical appraisal for project completion acceptance

1 Appraisal for reservoir impounding safety (introduction and key conclusions)

2 Technical appraisal for project completion acceptance (introduction and key conclusions)

L. 1. 9 Previous acceptances, remaining problems and treatment situation

L. 1. 10 Operation and management of the project

1 Management organization, staff and funds

2 Project handover

L. 1. 11 Initial operation and benefits of the project

1 Initial operation of the project

2 Benefits of initial operation

3 Analysis for observation and monitoring of the project

L. 1. 12 Compilation for financial accounts and audit

L. 1. 13 Problems and treatment comments

L. 1. 14 Plan for unfinished works

L. 1. 15 Experience and suggestions

L. 1. 16 Appendices

1 Statement for organizations and staff of the legal person

2 Approval documents and the adjusted ones of project proposal, feasibility study report and preliminary design

L. 2 Project Chronicle

L. 2. 1 According to the construction procedure of the small hydropower project, big influential events in the project development shall be recorded, including relevant official replies, instructions, major design changes, audit and examination by administrative departments, relevant agreements, major meetings, flood control, commencement and completion situation, and project acceptances and other important events.

L. 2. 2 The construction chronicle may be printed independently, but also can form the appendix to Working Report for Project Management.

L. 3 Working Report for Construction Management

L. 3. 1 Project overview

L. 3. 2 Bidding for the project

L. 3. 3 Management of construction progress

L. 3. 4 Major construction methods

L. 3. 5 Management of construction quality

L. 3. 6 Standardization construction and safe production

L. 3. 7 Contract management

L. 3. 8 Experience and suggestions

L. 3. 9 Appendices

1 Statement of organization and staff of construction management

2 Comparison between the planned resources and the ones actual used

3 Chronicle of construction management

4 List of technical standards used

L. 4 Working Report for Project Design

L. 4. 1 Project overview

L. 4. 2 Key points for project planning and design

L. 4. 3 Effect of design examination suggestions

L. 4. 4 Project criteria

L. 4. 5 Design changes

L. 4. 6 Quality management for design documents

L. 4. 7 Design service

L. 4. 8 Project evaluation

L. 4. 9 Experience and suggestions

L. 4. 10 Appendices

1 Statement of organization and staff of designer

2 Chronicle of project design

3 List of technical standards used

L.5 Working Report for Construction Supervision

L.5.1 Project overview

L.5.2 Supervision plan

L.5.3 Supervision process

L.5.4 Supervision effects

L.5.5 Project evaluation

L.5.6 Experience and suggestions

L.5.7 Appendices

1 Statement of organization and staff of supervisor

2 Chronicle of construction supervision

L.6 Working Report for Operation Management

L.6.1 Project overview

L.6.2 Operation management

L.6.3 Initial operation of the project

L.6.4 Data and analysis for project observation and monitoring

L.6.5 Experience and suggestions

L.6.6 Appendices

1 Official document for establishment of the management organization

2 Statement of setup and major staff of management organizations

3 List of rules and regulations for project operation

L.7 Report for Project Quality Supervision

L.7.1 Project overview

L.7.2 Quality supervision

L.7.3 Quality management system of participating organizations

L.7.4 Confirmation of project classification

L. 7. 5 Examination of project quality

L. 7. 6 Evaluation of project quality

L. 7. 7 Treatment of quality accidents and defects

L. 7. 8 Conclusions and suggestions for project quality

L. 7. 9 Appendices

1 Statement of quality supervisors for the project

2 Collection of quality supervision suggestions (written materials) during the construction

L. 8 Report for Project Safety Supervision

L. 8. 1 Project overview

L. 8. 2 Safety supervision

L. 8. 3 Safety management system of participating organizations

L. 8. 4 On-site supervision

L. 8. 5 Treatment of production safety accidents

L. 8. 6 Assessment of project safety production

L. 8. 7 Appendices

1 Statement of safety supervisors for the project

2 Collection of safety supervision suggestions (written materials) during the construction

Annex M Format of Working Report for Technical Pre-acceptance of Project Completion

×××Hydropower Project

Work Report for Technical Pre-acceptance of Project Completion

Technical pre-acceptance expert panel for completion of ××× hydropower project

Date

Foreword (including acceptance basis, organizations and process)

Section 1 Construction of the Project

1 Project overview

1.1 Name and location of the project

1.2 Major purposes and function of the project

1.3 Design content of the project

1.3.1 Proposal and design approval documents

1.3.2 Design criteria, scale and major technical and economic indices

1.3.3 Major contents and period of the construction

2 Construction process

2.1 Commencement and completion dates of major structures (table attached)

2.2 Key technical problems and the solutions

2.3 Key design changes

3 Completion situation and major completed quantities

4 Project acceptance and identification situation

4.1 Acceptances of individual unit construction

4.2 Stage acceptances

4.3 Special item acceptances (including major conclusions)

4.4 Technical appraisal for completion acceptance (including major conclusions)

5 Project quality

5.1 Project quality supervision

5.2 Project classification

5.3 Project quality inspection

5.4 Project quality assessment

6 Project operation and management

6.1 Management organizations, staff and fees

6.2 Project handover

7 Initial operation and benefits of the project

7.1 Initial operation of the project

7.2 Initial benefits of the project

7.3 Analysis for initial operation monitoring

8 Major problems and their treatment for previous acceptances and relevant appraisals

9 Construction plan for unfinished works

10 Assessment and suggestions

Section 2 Special Item (Works) and Acceptance

1 Land requisition compensation and resettlement of affected residents

1.1 Planning (design) situation

1.2 Completion situation

1.3 Acceptance situation and major conclusions

2 Water and land conservation facilities

2.1 Design situation

2.2 Completion situation

2.3 Acceptance situation and major conclusions

3 Environmental protection

3.1 Design situation

3.2 Completion situation

3.3 Acceptance situation and major conclusions

4 Project files (acceptance situation and major conclusions)

5 Fire control facilities (acceptance situation and major conclusions)

6 Others

Section 3 Financial Audit

1 Budget approval

2 Investment plan and fund usage

3 Investment completion and assets delivery

4 Funds for land requisition, relocation and resettlement of affected residents

5 Surplus funds

6 Estimated investment and expenses for uncompleted projects

7 Financial management

8 Completion financial report compiling

9 Audit and examination

10 Assessment and suggestions

Section 4 Comments and Suggestions

Section 5 Conclusions

Section 6 Signature List of Experts for Technical Pre-acceptance of Project Completion

Annex N　Format of Appraisal Report for Completion Acceptance

Completion Acceptance of ×××
Hydropower Project

Appraisal Report

Completion acceptance committee for
××× hydropower project

Date

Foreword (including acceptance basis, organizations and process)

1 Project design and completion situation

1.1 Name and location of the project

1.2 Major purposes and function of the project

1.3 Major contents of project design

1.3.1 Project proposal and design approval documents

1.3.2 Design criteria, scale and major technical economic indices

1.3.3 Major construction contents and period

1.3.4 Project investment and sources of the investment

1.4 Relevant organizations involved in the project (table attached)

1.5 Construction process

1.5.1 Commencement and completion dates of major structures

1.5.2 Key design changes

1.5.3 Key technical problems and their solutions

1.6 Completion situation and major completed quantities

1.7 Land requisition compensation and resettlement of affected residents

1.8 Soil and water conservation facilities

1.9 Environmental protection project

2 Project acceptance and appraisal

2.1 Acceptances of unit projects

2.2 Stage acceptances

2.3 Special item acceptances

2.4 Technical appraisal for completion acceptance

3 Major problems and treatment in previous acceptances and relevant appraisals

4 Project quality

4.1 Project quality supervision

4.2 Project item classification

4.3 Project quality tests (if any)

4.4 Project quality assessment

5 Budget execution situation

5.1 Investment plan and fund usage

5.2 Investment completion and assets delivery

5.3 Funds for land requisition and resettlement of affected residents

5.4 Surplus funds

5.5 Estimated investment for uncompleted projects and reserve funds

5.6 Completion financial report compiling

5.7 Audit

6 Plan for unfinished works construction

7 Project operation and management

7.1 Management organizations, staff and funds

7.2 Project handover

8 Initial operation and benefits of the project

8.1 Initial operation of the project

8.2 Initial benefits of the project

8.3 Analysis for initial operation monitoring

9 Technical pre-acceptance of completion project

10 Comments and suggestions

11 Conclusions

12 Reservations (with signature)

13 Signature list of members of committee and representatives from accepted organizations

14 Annex: Working report for technical pre-acceptance of completion project

Annex O Format of Warranty for Project Quality

×××Hydropower Project

Warranty

Contractor:

Date

Warranty of ××× hydropower project

1 Acceptance of contract project completion

2 Scope and contents of the warranty

3 Warranty period

4 Warranty duty

5 Warranty fees

6 Others

Contractor：

Legal representative：(signature)

Date：

Annex P Format of Completion Certificate for Contracted Project

××× Hydropower Project

××× Contracted Project

(Contract name and serial number)

Completion Certificate

Legal person:

Date

Legal person:

Construction agent (if any):

Designer:

Supervisor:

Contractor:

Major equipment manufacturer (supplier):

Operation and management organization:

Completion Certificate of Contracted Project

××× contracted project of ××× hydropower station has passed the completion acceptance chaired by ××× on ××.××.×××× (date), and now the Completion Certificate of Contracted Project is issued.

Legal person:

Legal representative: (signature)

Date:

Annex Q　Format of Warranty Termination Certificate for Project Quality

×××Hydropower Project

(Contract name and serial number)

Warranty Termination Certificate

Legal person:

Date

××× Hydropower Project

Warranty Termination Certificate

The quality warranty of ××× hydropower project (contract name and serial number) expired on ××. ××. ×××× (date) and the responsibility in the contract has been fully executed. Now the Warranty Termination Certificate is issued.

Legal person:

Legal representative: (signature)

Date:

Annex R Format of Completion Certificate (Original)

Completion Certificate of ××× Hydropower Project
××× hydropower station has passed the completion acceptance chaired by ××× on ××. ××. ×××× and now the Completion Certificate is issued.
Authority organization:
Date:

Note: the cover size of the original certificate: length 60cm×width 40cm.

Annex S Format of Completion Certificate (Copy)

×××Hydropower Project

Completion Certificate

Date

Presiding organization:

Supervision and management organization for legal person acceptance:

Legal person:

Construction agent (if any):

Designer:

Supervisor:

Major contractor:

Major equipment manufacturer (supplier):

Supervision organization for quality and safety:

Operation and management organization:

Commencement date:

Completion date:

Completion Certificate of ×××
Hydropower Station

××× hydropower station has passed the completion acceptance chaired by ××× on ××. ××. ×××× (date), and now the Completion Certificate is issued.

Authority organization:

Date:

Explanation of Wording

Words in this code	Same expressions in special cases	Different degrees of strictness
Shall	It is necessary	Required
Shall not	It is not allowed/permitted It is unacceptable	
Should	It is recommended It is advisable	Recommended
Should not	It is not recommended It is not advisable	
May	It is suitable It is desirable It is preferable	Permitted
Need not	It is unnecessary It is not required